discovermore

Earth's Processes

What Is the Water Cycle?

Doug Bradley

AF898707

Britannica Educational Publishing

IN ASSOCIATION WITH

THE ROSEN PUBLISHING GROUP

Published in 2027 by Britannica Educational Publishing (a trademark of Encyclopædia Britannica, Inc.) in association with The Rosen Publishing Group, Inc.
2544 Clinton Street, Buffalo, NY 14224

Copyright © 2027 by Encyclopædia Britannica, Inc. Britannica, Encyclopædia Britannica, and the Thistle logo are registered trademarks Encyclopædia Britannica, Inc. All rights reserved.

Rosen Publishing materials copyright © 2027 The Rosen Publishing Group, Inc. All rights reserved.

Distributed exclusively by Rosen Publishing.
To see additional Britannica Educational Publishing titles, go to rosenpublishing.com.

All rights reserved. No part of this book may be reproduced in any form without permission in writing from the publisher, except by a reviewer.

Editor: Greg Roza
Book Design: Deanna Lepovich

Photo Credits: Cover Karen Benjamin/Shutterstock.com; (series background) Dai Yim/Shutterstock.com; p. 4 The Syndicate/Alamy Photos; p. 5 Natasha Zakharova/Alamy Photos; p. 6 Imago/Alamy Photos; p. 7 Papia Majumder/Shutterstock.com; p. 8 Alex Hinds/Alamy Photos; p. 9 mauritius images GmbH/Alamy Photos; p. 11 (top) John Lander/Alamy Photos; p. 11 (bottom) Levente Bodo/Alamy Photos; p. 12 Tom Thulen/Alamy Photos; p. 13 Mario Beauregard Beaustock/Alamy Photos; p. 14 denayunebgt/Shutterstock.com; p. 15 David R. Frazier Photolibrary, Inc./Alamy Photos; p. 17 (top) Kryuchka Yaroslav/Alamy Photos; p. 17 (bottom) Namita Mitra/Alamy Photos; p. 18 Dimitrios Karamitros/Shutterstock.com; p. 19 YAY Media AS/Alamy Photos; p. 20 Gino Rigucci/Alamy Photos; p. 21 Ralph Rozema/Alamy Photos; p. 22 Khatawut Chaemchamras/Alamy Photos; p. 23 Pete Niesen/Alamy Photos; p. 24 Balthasar Thomass/Alamy Photos; p. 25 Edwin Remsberg/Alamy Photos; p. 26 David R. Frazier Photolibrary, Inc./Alamy Photos; p. 27 Mamunur Rashid/Alamy Photos; p. 28 CACIO DE VASCONCELOS/Alamy Photos; p. 29 Dmytro Flisak/Alamy Photos.

Library of Congress Cataloging-in-Publication Data

Names: Bradley, Doug, 1971- author
Title: What is the water cycle? / Doug Bradley.
Description: [Buffalo] : Britannica Educational Publishing, [2027] |
Series: Discover more: Earth's processes | Includes index.
Identifiers: LCCN 2026005341 (print) | LCCN 2026005342 (ebook) | ISBN 9781641906234 library binding | ISBN 9781641906227 paperback | ISBN 9781641906241 ebook
Subjects: LCSH: Hydrologic cycle--Juvenile literature | Water--Juvenile literature
Classification: LCC GB848 .B73 2027 (print) | LCC GB848 (ebook)
LC record available at https://lccn.loc.gov/2026005341
LC ebook record available at https://lccn.loc.gov/2026005342

Printed in China

Some of the images in this book illustrate individuals who are models. The depictions do not imply actual situations or events.

CPSIA Compliance Information: Batch #QSBRIT27. For further information contact Rosen Publishing at 1-800-237-9932.

Find us on

Contents

The Blue Planet

Nearly three-fourths of Earth's surface is covered with water. All plants and animals need water to live. Water exists in many different forms, such as salt water in oceans and fresh water in rivers or lakes. Other examples include frozen water of glaciers or water found deep in the ground. The amount of water on the planet remains the same, but it constantly moves around. The way water moves around Earth is called the water cycle.

Earth is sometimes called the blue planet because its surface is largely covered with water, making it look blue from outer space.

We use water for so many things every day.

People have many uses for water besides drinking. They use it for washing and cooking. They use it to irrigate crops and lawns, to clean streets, and to operate air-conditioning units and heating systems. Flowing water is used to produce electricity.

Consider This

How would your life be different if water wasn't as plentiful as it is?

States of Matter

A molecule is the smallest particle of a substance having all the characteristics of the substance. They are combinations of even smaller pieces of matter called **atoms**. A molecule of water is made of two hydrogen atoms and one oxygen atom. Water makes up about 60 percent of an adult's body weight. Water helps the body get rid of waste and maintain its temperature.

This is a visual representation of a single water molecule. The scientific formula for water is H_2O.

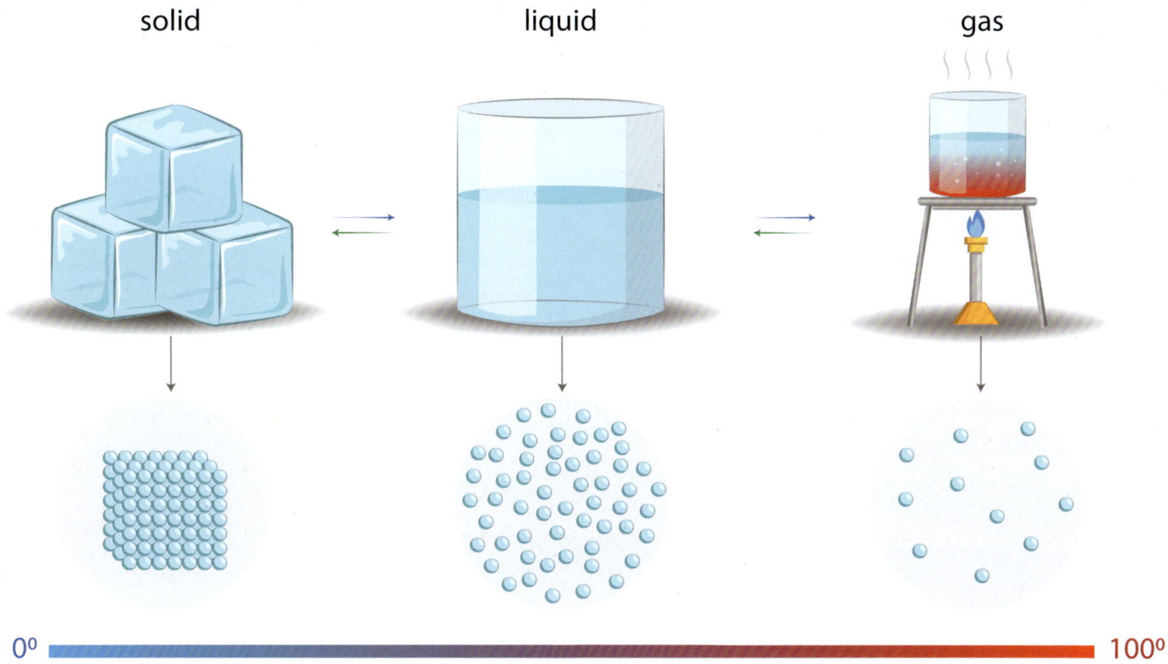

As a substance increases in temperature, the spaces between its atoms increases.

Water can be found in three physical states: liquid, solid (ice), or gas (steam or vapor). The molecules in all three states are constantly moving. In snow or ice, the water molecules vibrate but basically stay in place. In rain, the molecules move more quickly but stay near each other. In steam, the molecules move so quickly that they fly away in all directions.

WORD WISE

ATOMS ARE TINY PARTICLES–SO TINY WE CAN'T SEE THEM! ATOMS ARE THE BUILDING BLOCKS OF ALL MATTER.

Heating and Cooling

When ice is heated, the water molecules move faster and farther apart. This melts the ice into liquid water. When liquid water is heated, the molecules speed up even more. Molecules at the surface begin to break loose and fly into the air. At most places on Earth, fresh water boils when it reaches 212 °F (100 °C). Once it reaches this temperature, water cannot get any hotter.

People use boiling water to cook foods such as eggs, pasta, and vegetables.

When the weather turns colder, many lakes freeze over. Ice fishers wait for the ice on lakes to be thick enough to cut a hole in it.

Very cold temperatures turn water into ice. Fresh water freezes at 32 °F (0 °C). Water is lighter as a solid than as a liquid. This is why ice floats and why it forms on top of a lake instead of at the bottom. Water also expands when it freezes. This is why water pipes can burst on cold nights. The water inside the pipes expands as it freezes until the ice cracks the pipes.

Consider This

Glaciers cover about 30,000 square miles (77,700 square kilometers) of the United States, most of which is in the state of Alaska. What would happen if the glaciers melted?

Water Vapor

Boiling is not the only way liquid water can become water vapor. At temperatures below boiling, liquid water becomes vapor through a process called evaporation. When the sun warms the water at the surface of a puddle, the water evaporates into invisible gas. The water seems to disappear, but it actually moves into the air as a gas called water vapor. The main source of water vapor in Earth's atmosphere is the surface water of the planet's oceans.

When a person exhales, their breath has water molecules in the form of vapor. Plants also release water vapor into the air through a similar process called transpiration. Water is drawn through the roots of a plant to its leaves, from where it evaporates. Leaves have tiny holes in them that release water vapor.

Salt farms collect the salt leftover after the sun has evaporated salt water.

compare and contrast

How are breathing and transpiration similar? How are they different?

The tiny holes on a plant's leaves that release water vapor are called stomata. They are far too small to see without a microscope.

Condensation

Condensation is the opposite of evaporation. As water molecules in vapor cool down, they lose heat energy and slow down. The slower water molecules collect together to form a liquid. Condensation happens when there is so much water vapor in the air that the air can no longer hold the moisture. Near the ground, water vapor condenses to become fog. Up in the sky, it forms clouds.

Fog can form over a warm body of water when very cold air moves over it. This is sometimes called sea smoke.

Dew forms when warm air passes over cool surfaces. Dew is common in the morning as the sun rises and heats the cold Earth.

Another example of condensation is the drops of water that form on the outside of a glass of a cold beverage. The drops seem to appear from nowhere, but they actually form from water vapor in the air. The dew that forms on grass overnight is an example of condensation, too. At dawn, tiny drops of water often cling to plants and grass, creating dew.

Consider This

Too much condensation can be harmful. In what ways might too much moisture cause problems for farmers or homeowners?

The Way Water Moves

When liquid water from oceans and seas evaporates, it forms clouds in the sky. The wind pushes these clouds over land, where the water vapor condenses and becomes rain or snow. The rain or snow falls back to the ground. Water has been cycling since Earth formed. The water in your cup may have been sipped by dinosaurs millions of years ago!

Rain and snow are both forms of precipitation. So are sleet and hail.

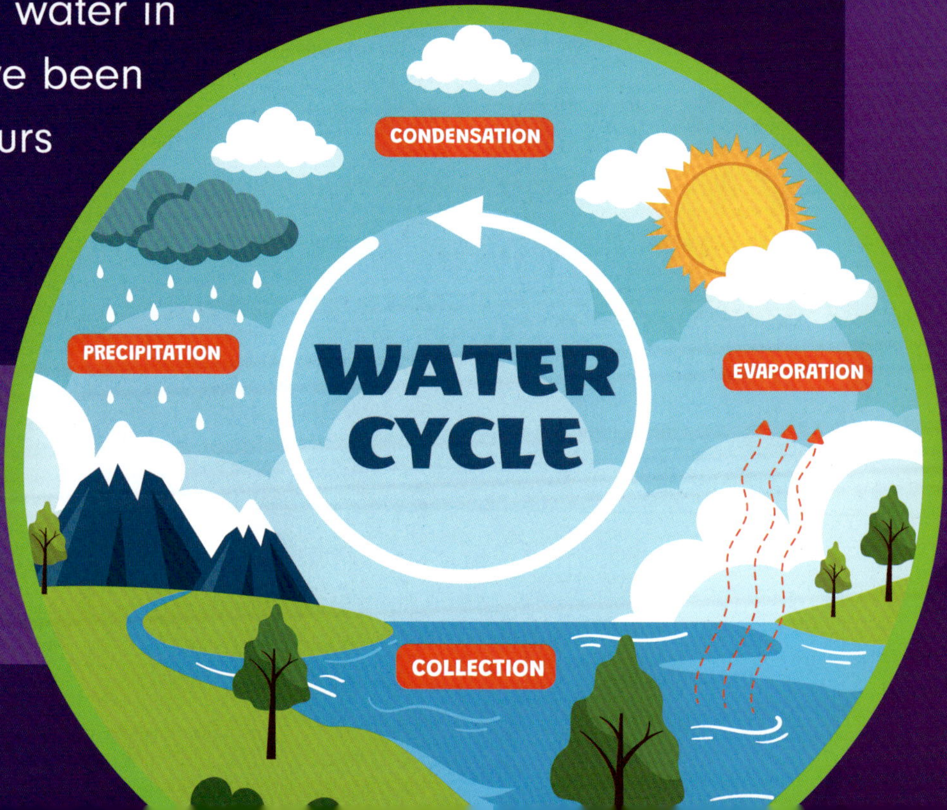

Sometimes the ground becomes so full of water it can't hold anymore. The extra water flows over the ground into streams and rivers. This is called runoff.

There are three main steps involved in the circular path of the water cycle: evaporation, condensation, and precipitation. Water evaporates into the air from rivers, lakes, plants, and oceans. The evaporated water condenses into dew, **frost**, fog, and clouds. Precipitation falls to Earth from these clouds. The most common forms of precipitation are rain and snow. Some of the rainwater or melted snow joins rivers and flows down to the ocean or sea. Then the cycle starts all over again.

WORD WISE

FROST IS A THIN LAYER OF ICE CRYSTALS THAT FORMS ON A COLD SURFACE.

Precipitation

The air high up in the sky is full of invisible water vapor that condenses into clouds. A cloud is made up of millions of tiny water droplets or ice crystals floating together in the air. When water vapor cools, or the air can no longer hold the amount of water vapor it contains, the water returns to a liquid state and falls as precipitation.

Precipitation may come in many forms, including rain, hail, sleet, and snow. The form precipitation takes depends on other weather conditions, such as temperature. Precipitation size, speed, and rate are different all over the world. In the deserts of Egypt it may rain less than 1 inch (2.5 centimeters) per year. However, on June 17, 2022, the Indian town of Mawsynram received 39.5 inches (100.3 cm) of rain in 24 hours!

Clouds are classified mainly by their appearance. Not all clouds are a sign of precipitation.

compare and contrast

How are rain, snow, sleet, and hail similar to each other? How are they different?

Mawsynram, India, is the rainiest place on Earth. This tiny town received about 1,024 inches (2,600 cm) of rain in 1985 alone!

Around We Go!

Rain and snow collect in different places, such as seas, oceans, rivers, or lakes. The main sources of lake water include precipitation (rain or snow), rivers and streams, and melting ice and snow. Some water is absorbed by plants and eventually evaporates, while other precipitation seeps into the ground. This water below Earth's surface is called groundwater. It can reach lakes through openings called springs. Groundwater is a major source of fresh water. By means of wells, humans bring this water to the surface to satisfy their need for water.

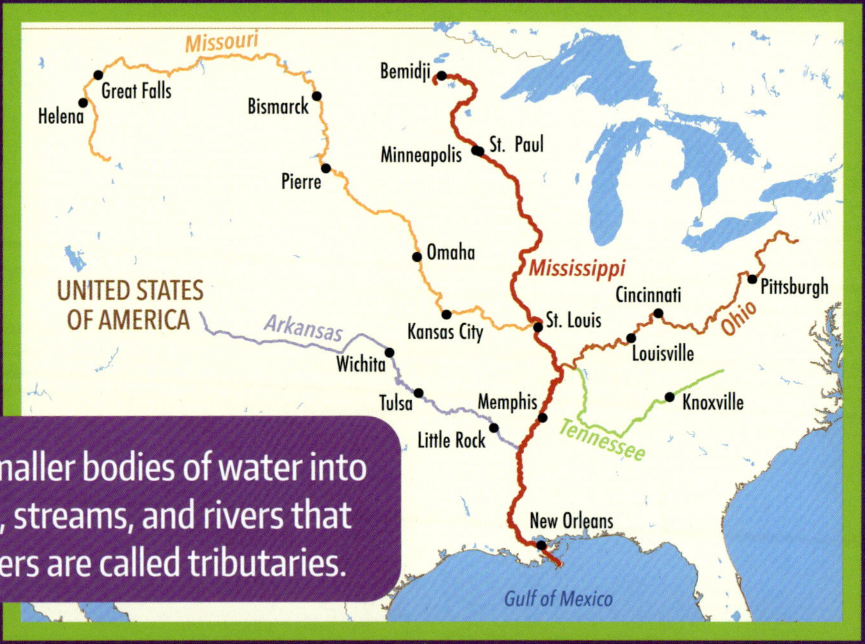

Water flows from smaller bodies of water into larger ones. Creeks, streams, and rivers that flow into larger rivers are called tributaries.

Snow on cold mountains builds up over the years. The weight of newer snow packs the lower levels tightly. In time this forms a glacier.

Snowcapped mountains also collect water in the form of snow. As the snow or ice melts with warm weather, the water trickles down the mountain and into streams or rivers. The water is warmed by the sun and enters the atmosphere again through evaporation, and the water cycle continues.

Consider This

Water carves and shapes Earth's surface, wearing away rocks. What examples of this have you seen?

What Is Climate?

Climate is the weather found in a certain place over a long period of time. Many factors affect climate, including the sun, oceans, winds, land types, clouds, and human activities. Clouds cool and warm Earth's surface. They work like a blanket and trap the heat close to Earth. They also keep Earth cool as sunlight bounces off of them back into space. The ocean also warms or cools Earth's surface. It warms land in winter and cools it in summer because water cools and heats more slowly than land.

Oceans affect the climate of places along coasts.

Warmer global temperatures are melting polar ice, putting more liquid water in the water cycle.

The average surface temperature on Earth is slowly increasing. This trend is known as **climate change**. It affects the water cycle. Warmer temperatures increase evaporation. This results in drought in some areas and too much precipitation in other areas. In places with warmer winters, precipitation falls as rain and not as snow. Warmer temperatures cause glaciers to melt.

WORD WISE

CLIMATE CHANGE IS A CHANGE IN THE PATTERNS OF WEATHER FOUND OVER LONG PERIODS OF TIME ON EARTH.

The Human Water Cycle

The human water cycle explains how people collect, use, and reuse water. Humans need water for daily tasks such as drinking, washing, and waste removal. Some people get water from a river or well. In cities and towns, water supply systems provide water to residents and businesses. These systems collect, deliver, purify, store, and distribute the water.

Water makes up more than half of the weight of the human body.

Modern agriculture often requires chemicals for fertilizers and pesticides. Runoff from large farms often pollutes nearby waterways.

What happens to water after it's flushed down the toilet or rinsed down the drain? It is taken to a wastewater treatment center. There, workers clean it up before delivering it back into the environment and into the water cycle. The workers remove contaminants from the flushed water, such as food scraps, soap chemicals, and human waste. Wastewater treatment centers also clean up storm water from streets.

People do not always have easy access to clean, drinkable water. It is a great effort to keep water clean and available. This is a serious problem in many parts of the world.

Consider This

Less than 1 percent of Earth's fresh water is available for people to use. How can people keep from wasting water?

Energy is another part of the human water cycle. Power plants burn fuels to boil water into steam. The steam powers a machine that produces electricity. Water is also used during the cooling process in power plants. A water power plant uses the moving mechanical energy of flowing river water to produce electricity. The up-and-down motion of waves in oceans, or tidal energy, can also be used to create **hydroelectricity**.

The Three Gorges Dam in China is the largest hydroelectric plant in the world.

Scientists are working to find new ways to irrigate crops to keep our water supplies safe.

People also use the water cycle for the production of food. Irrigation is what farmers do when they add water to their fields. The water makes their crops grow. Irrigation takes the place of rainfall in dry regions. However, irrigation can also cause problems with the environment. Taking too much water to use for irrigation can cause water shortages in other places. Also, irrigation water contains more salt than rainwater. Salt can build up in the soil and harm plants.

WORD WISE

HYDROELECTRICITY IS ELECTRICITY THAT HAS BEEN PRODUCED USING THE POWER OF MOVING WATER.

Harming the Water Cycle

Deforestation is the clearing of forestland for building projects and agriculture. When trees are cut down, less water enters the atmosphere through the leaves of trees. In turn, there is less rain. The region may experience drought and other problems. Land with fewer plants growing on it is less able to absorb water. Rainwater washes away quickly instead of being stored in plants and returned slowly to the atmosphere.

Deforestation of the Amazon Rainforest, both legal and illegal, is a hotly debated topic in Brazil and around the world.

Industrial waste has negative effects on all living things and the environment.

Pollution affects the water cycle. People dump garbage and sewage into creeks, rivers, ponds, lakes, and oceans. Factories or cities sometimes release oils, poisonous chemicals, and other wastes into water. Chemicals in fertilizers or pesticides can seep into the ground and make the groundwater unfit to drink. Pollution does not just affect the creatures that live in the water. Fish taken from polluted streams pass the pollution into the bodies of people who eat them.

Consider This

What can people do to counter the effects of deforestation?

What Can We Do?

As Earth's population grows, water conservation has become a necessity. Conservation is the protection of things found in nature. It requires the sensible use of all Earth's natural resources, including water. People need to be careful about how much water they use and be sure not to waste any. Water conservation can be as simple as turning off the water faucet while brushing your teeth or taking shorter showers.

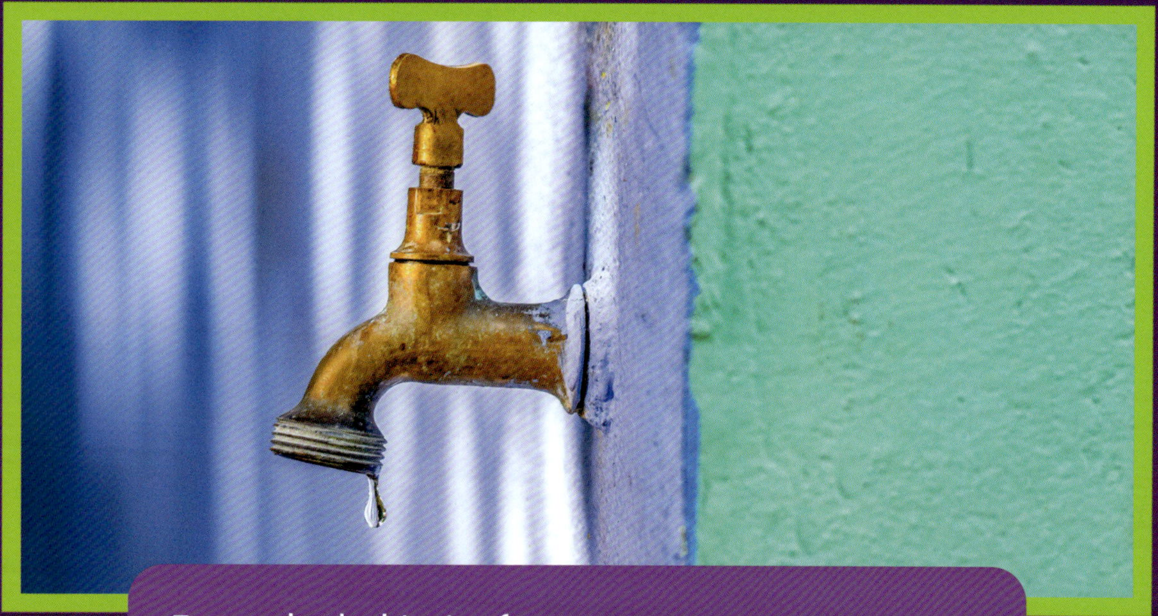

Even a slowly dripping faucet can waste water over time.

You can help to protect the water cycle by volunteering to help clean up polluted areas. Volunteer means to do something without being paid to do it.

People who care about conservation also try to keep the environment clean and healthy, including lakes, streams, and oceans. Planting trees near riverbanks helps to keep water clean. Keeping toxic cleaners and domestic animal waste from water also helps. We can all be responsible stewards of the environment by conserving water and keeping it clean.

Consider This

What can you do to help the water cycle? What can you do at home? What can you do outside your home?

Glossary

atmosphere The layer of gases that surrounds Earth.
drought A long period of dry weather.
environment All the physical surroundings on Earth, including all living and nonliving things.
fertilizer A substance used to help crops grow.
fog Fine particles of water floating in the atmosphere near the ground.
gas A fluid (such as air) that has no fixed shape and tends to expand without limit.
glacier A large body of ice moving slowly down a slope or valley or spreading outward on a land surface.
hail Small lumps of ice that fall from clouds sometimes during thunderstorms.
irrigate To supply with water by artificial means.
mechanical energy All the energy that an object has because of its motion and its position.
natural resource Something found in nature that can be used by people, such as sunlight, plants, or water.
pesticide A substance used to kill pests that would harm crops or that would keep the crops from growing.
pollution Dirtied by waste, chemicals, or other harmful substances.
precipitation Water that falls to Earth as hail, mist, rain, sleet, or snow.
recycle To remake something so it can be used again.
sleet Frozen or mostly frozen raindrops.
steam The invisible vapor into which water is changed when heated to the boiling point.
steward Someone who takes care of people or places.

For More Information

Books

James, Buckley, Jr. *The Perito Moreno Glacier of Argentina.* Bloomington, MN: Bearport Publishing, 2026.

James, India. *Ocean Engineering: Protecting Our Ocean Environments.* Minneapolis, MN: Learner Publications, 2026.

Websites

Climate Change

kids.nationalgeographic.com/science/article/climate-change

National Geographic Kids offers a wealth of information regarding climate change and what we can do about it.

The Water Cycle

www.noaa.gov/education/resource-collections/freshwater/water-cycle

Learn more about the water cycle from the National Oceanic and Atmospheric Administration.

Publisher's note to educators and parents: Our editors have carefully reviewed these websites to ensure that they are suitable for students. Many websites change frequently, however, and we cannot guarantee that a site's future contents will continue to meet our high standards of quality and educational value. Be advised that students should be closely supervised whenever they access the internet.

Index